Children's Environmental Awareness

**For all children who wish
to take care of Environment**

VOL.1 OF 5

RECYCLING

Jahangir Asadi

Vancouver, BC CANADA

Copyright © 2022 by **SILOSA** Consulting Group Inc.

Published by: Silosa Consulting Group Inc.
Vancouver, BC **CANADA**
Email: Info@Silosa.ca
www.silosa.ca

Ordering Information:
Quantity sales. Special discounts are available on quantity purchases by universities, schools, corporations, associations, and others. For details, contact the "Sales Department" at the above mentioned email address.

Children's Environmental Awareness Vol.1/J.Asadi —1st ed.
ISBN: 978-1-990451-49-2

Contents

We hope that, 10,000 years from now, future generations will be able to see flowers that provide bees with nectar and pollen and...
BEES provide flowers with the means to reproduce by spreading pollen from flower to flower,....

J.Asadi

This book is dedicated to my professor, Dr.Sadeq Fakhr

The more you care about our environment, the more it will be protected from contaminants and toxins

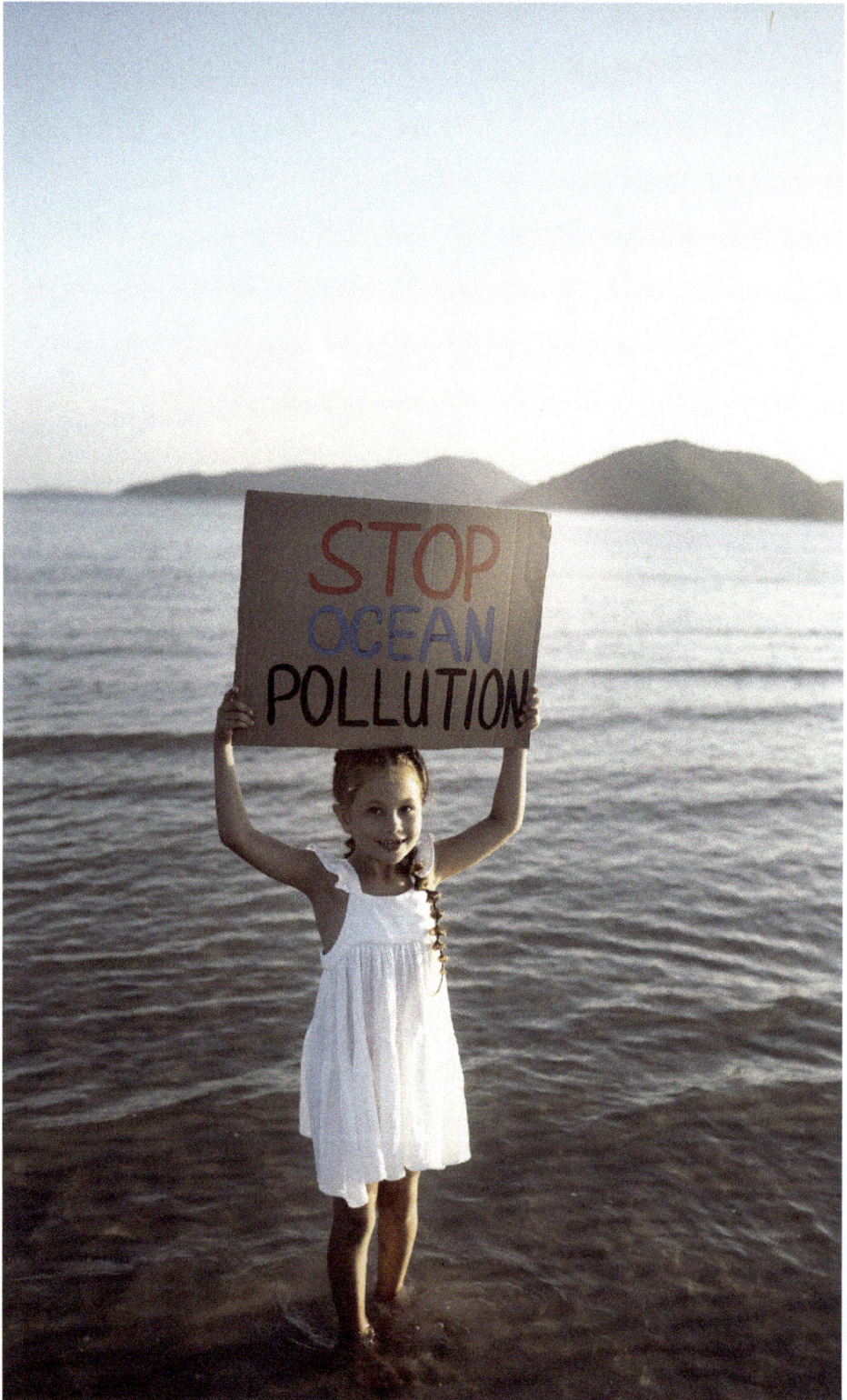

SILOSA Consulting Group (SCG)

Silosa Consulting Group (SCG) was established to provide outstanding consulting services of management system standards to individuals, groups, companies, and organizations all over the globe.

SCG is publishing an "Children's Environmental Awareness" book series related to increasing environmental awareness of kids means being aware of the natural environment and making choices that benefit the earth, rather than hurt it. Vol.1 to 5 providing some of the ways to practice environmental awareness include: **Recycling, Conserving energy and water, Reuse, Activism, and others**.

SCG book publishing services and distribution services are connected to over 39,000 booksellers worldwide, including Apple, Amazon, Barnes & Noble, Indigo, Google Play Books, and many more.

SCG has enough experiences to help create new and effective programmes in different countries all over the world. For more detail, visit our website : http://silosa.ca and/or send your enquiery to the following email:

info@silosa.ca

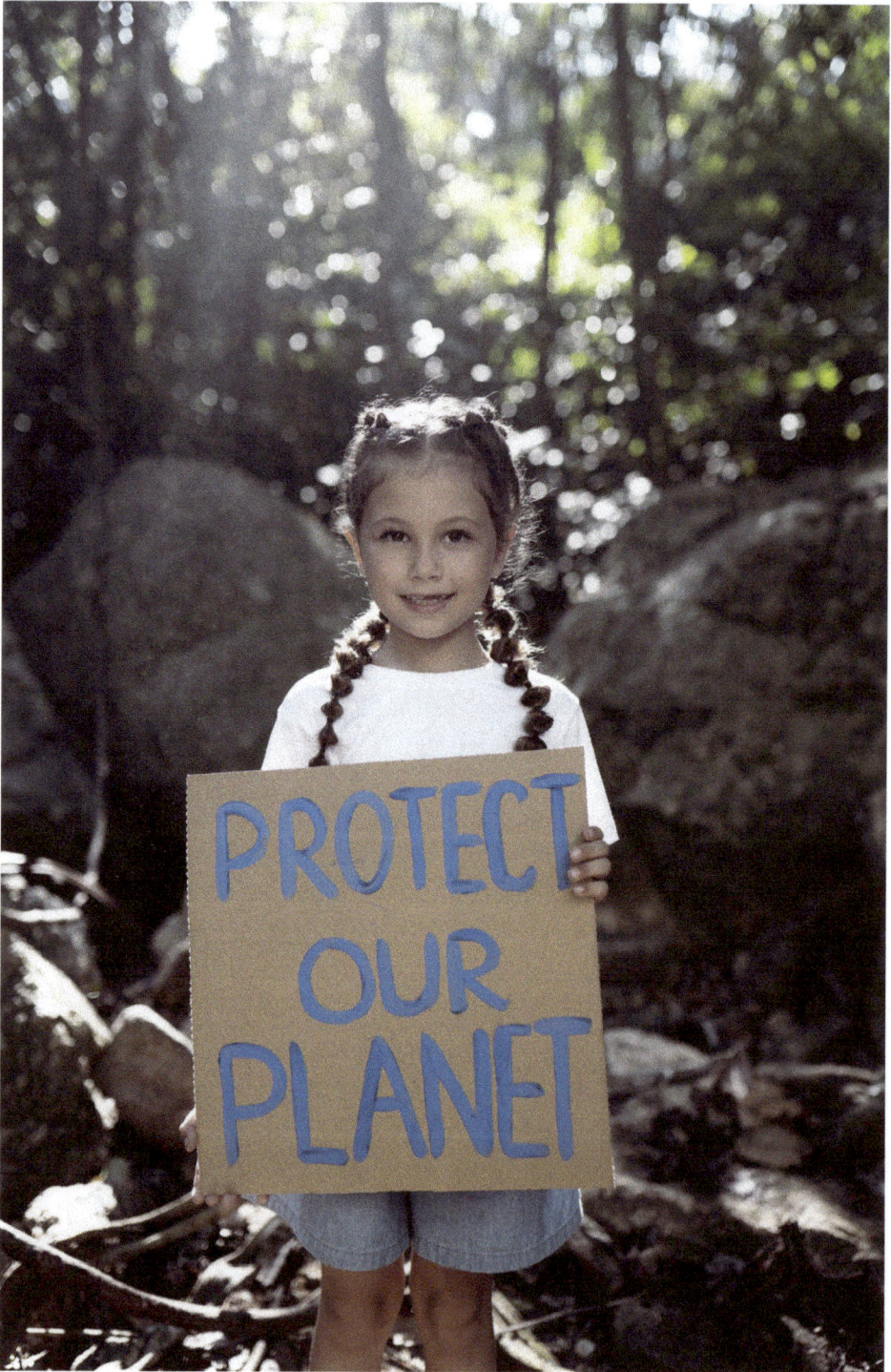

CHAPTER 1

About ISO 14000 for Kids

The International Organization for Standardization is an independent, non-governmental organization, the members of which are the standards organizations of the 165 member countries. It is the world's largest developer of voluntary international standards and it facilitates world trade by providing common standards among nations. More than twenty thousand standards have been set, covering everything from manufactured products and technology to food safety, agriculture, and healthcare.

Kids ISO 14000s

"Kids ISO 14000s" is a new environmental education program for children, based on ISO 14000s, which is international standard for environmental management. Primary aims of this program are: -

1. To teach and train children how to manage the environmental issues (such as energy saving) by themselves through the working book and guide book of this program,

2. To certify those children who showed good accomplishment in the program from highly international authority (as is the case of ISO 14000s)

3. To network those children through the international network (Kids International Network), so that the children can work on the environment, internationally.

2. System of Kids ISO 14000s Program
The system of Kids ISO 14000s Program consists of
1. Operation Headquarter (ArTech).
2. Workbook, Guidebook (originally published by ArTech, and local versions are produced by each countries).
3. Eco-Kids-Instructors for local operation and evaluation of the performance of the children.
4. International accreditation committee for accreditation of accomplishment of the children, for certification of the Eco-Kids-Instructors, as well as overall checks of this program.
5. Linkage with international organizations (such as UNU, UNESCO, etc. …) And also national organizations

More information can be obtained :

www.ISO.org

Canada

Environmental Sustain for Future kids established in Vancouver, BC Canada in 2020. (ESFK) is an international ecolabel focused on taking care of environment for future of kids.

ESFK defined as 'self-declared' environmental claims made by manufacturers and businesses based on ISO 14020 series of standards, the claimant can declare the environmental objectives and targets in relation to taking care of environment for future kids. However, this declaration will be verifiable.

Environmental Sustain for Future Kids
Vancouver, BC CANADA

Email: info@esfk.org
Web: www.esfk.org

CHAPTER 3

The Environment and Kids

1 Teach children what can be recycled instead of becoming household waste.

2 Have kids put old homework in the paper bin & yogurt containers in the plastic bin.

3 "Complete the circle" by buying products made with recycled ingredients such as paper.

12

ACTIVITIES FOR KIDS
TO LEARN ABOUT
RECYC

1: Build different toys with cardboard, don't throw away the cardboard boxes and food containers!

2: Play a game to help kids practice which items can be recycled, and which are waste.

3:Make a bird feeder, This is one of those activities that demonstrates how much cheaper it is to use recycled materials than to buy something brand new.

4:Paint the symbol, Practice recognizing this symbol through art. This way, kids can easily identify recycling bins out in public.

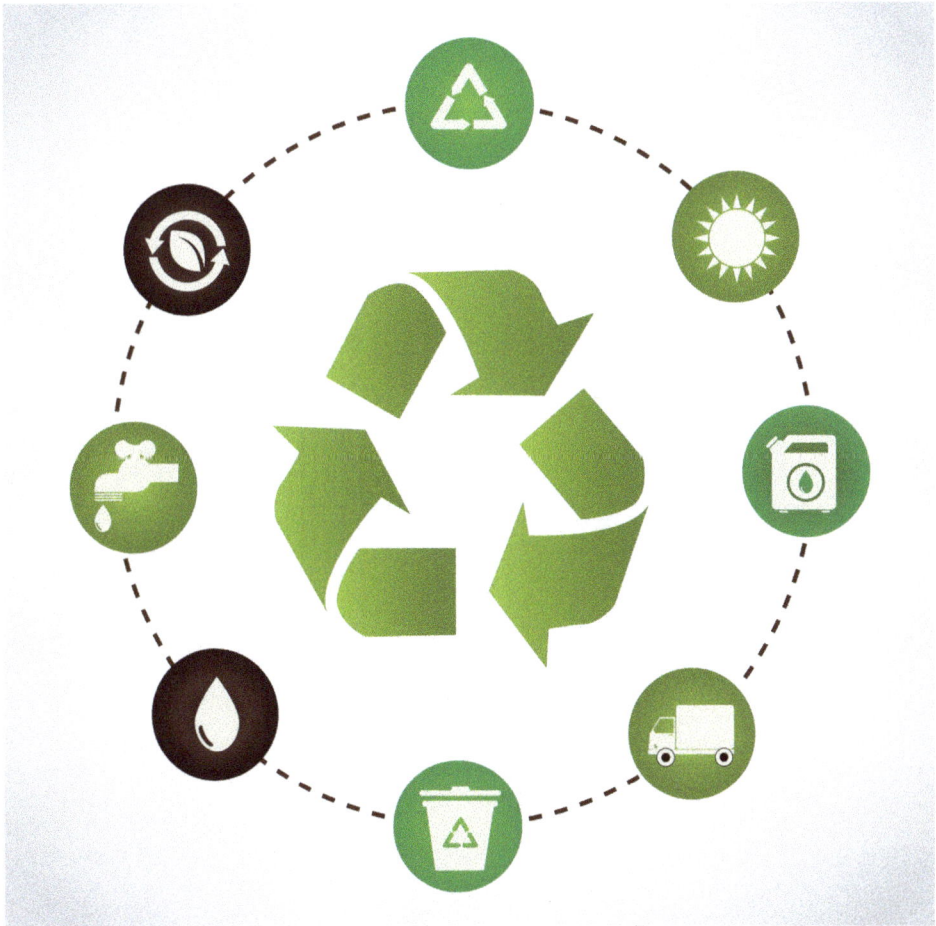

5:Making small bins that can be kept indoors allows kids the chance to easily sort the Garbage.

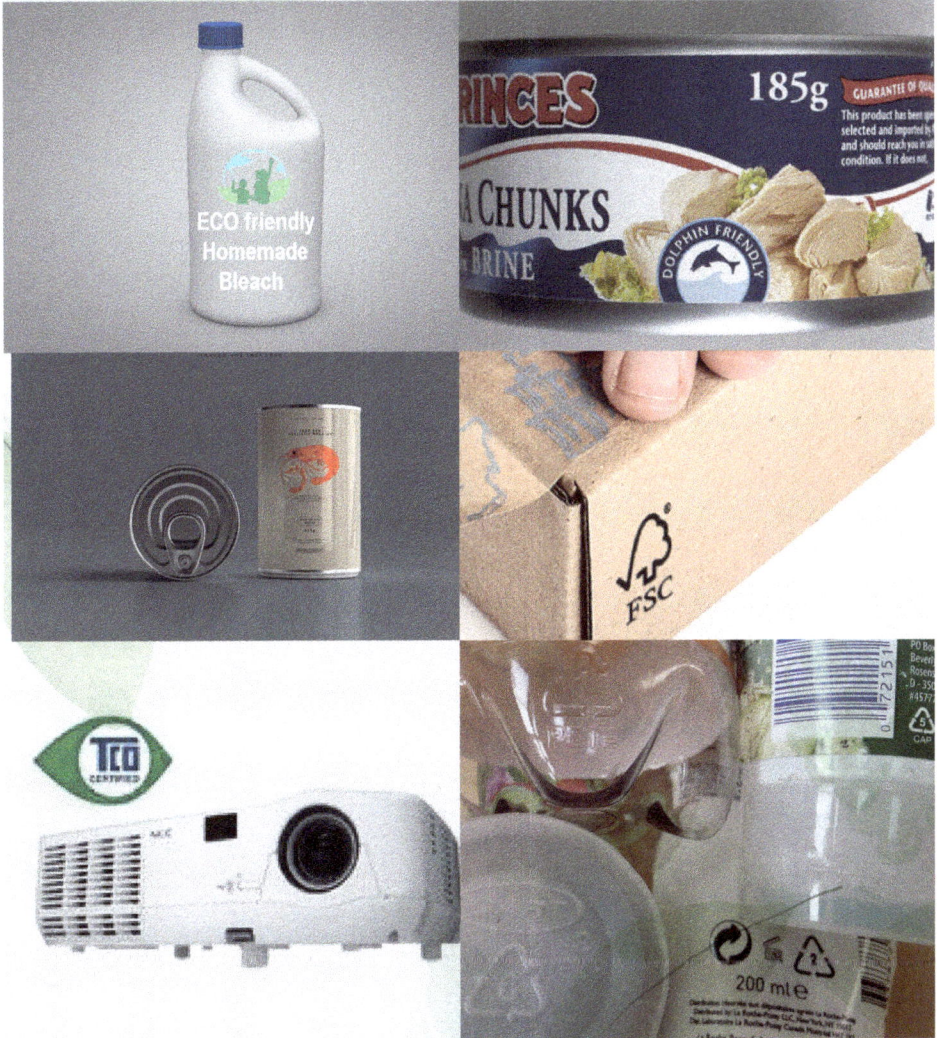

6: Finding Ecolabels on products during shopping: FSC on copy papers and notebooks, Recyclable & ESFK on many products, Dolphin safe and Salmon safe on Tuna fish cans...

7:Homemade puzzles, Instead of throwing away greeting cards, cut them up and make a homemade puzzle.

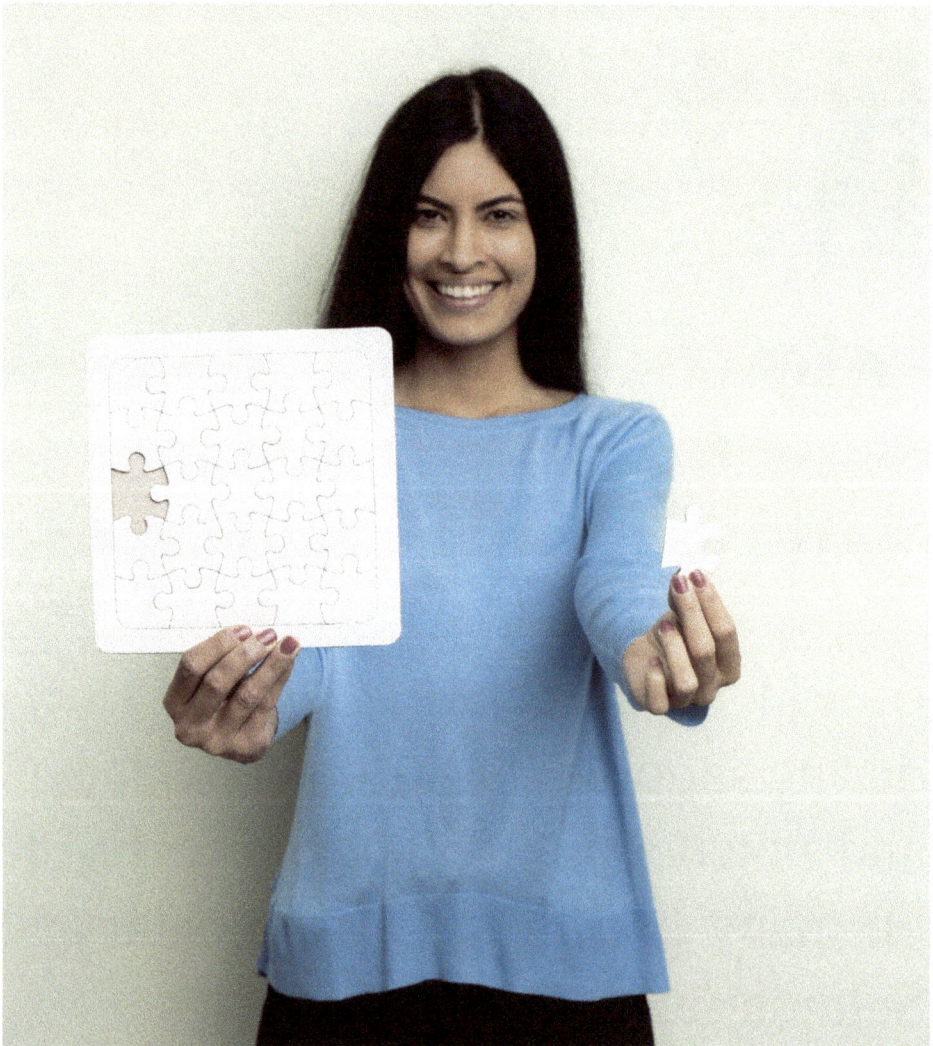

8:Take a field trip!
Head out to a local recycling center and watch a bit of the process happen.

9:Spearhead a recycling club. Help your child start a local initiative in their neighbourhood or school.

10: Donate toys and clothes.

Instead of throwing away toys your kids are done playing with, help them choose some to be donated and reused by other children.

11:Make a DIY toy, Before recycling items, try to re purpose them. A milk jug or a yogurt container can make for really fun ball poppers for kids.

12:Pack a waste-free lunch. Use a recycled container in lieu of items in bags. More ideas for a waste-free lunch

Photo

Gallery

TO LEARN ABOUT

RECYC

CLING

BIBLIOGRAPHY

Bibliography:

Asadi, J., " International Environmental Labelling Book Series Vol.1-11, 2022

Asadi, J., "International Environmental Labelling, Economic Consequences, Export Magazine, July 2001

Asadi, J. 2008. Mobile Phone as management systems tools, ISO Magazine, Vol.8, No.1

Asadi, J., Eco-Labelling Standards, National Standard Magazine, Sep. 2004.

Birett, M. J. 1997. Encouraging Green Procurement Practices in Business: A Canadian Case Study in Program Development (108-118). in Greener Purchasing : Opportunities and Innovation. Sheffield, Greenleaf Publishing 325p.

Bowen, Nicola, World Agrochemical Markets, PJB Publications Ltd., March 1991.

Burnside, A., (1990), Keen on Green, Marketing, 17 May, pp35-36

Butler, D., (1990), A Deeper Shade of Green, Management Today, June, pp74-79

Cairncross, F. 1995. Green, Inc.: A guide to business and the environment. London, Earthscan. 277p.

ISO 14000 for kids program, ISO, ISO.org

ISO (2015). ISO Survey 2015 (online). Retrieved from: http//www.iso.org.

ISO (2018). ISO 19011 - Guidelines for auditing management systems quality management systems. Geneva: International Organization for Standardization.

ISO (2019). ISO 9000 Family - Quality Management. Retrieved from: https://www.iso.org/home.html.

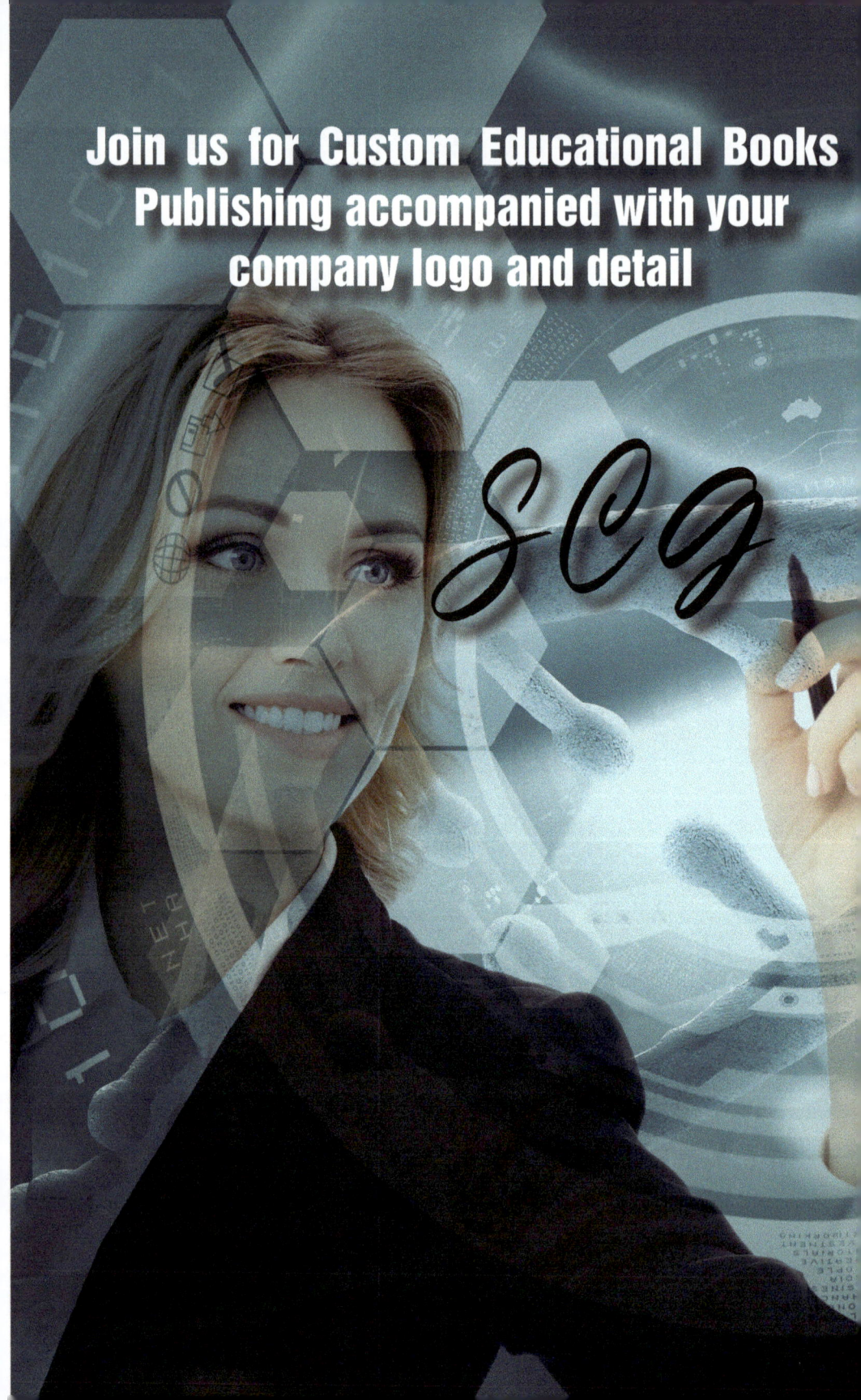

Join us for Custom Educational Books Publishing accompanied with your company logo and detail

SCG

Children's
Environmental
Awareness
Vol.1 of 5 : Recycling

Environmental Sustain for Future Kids

Children's
Environmental
Awareness
Vol.2 of 5 : Greener Energy

Environmental Sustain for Future Kids

Children's
Environmental
Awareness
Vol.3 of 5 : Reuse

Environmental Sustain for Future Kids

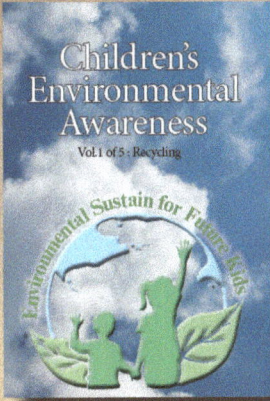

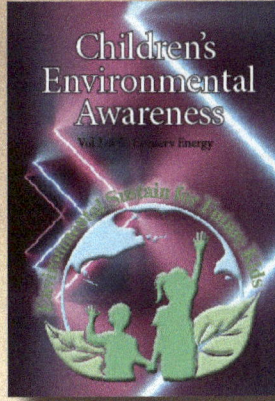

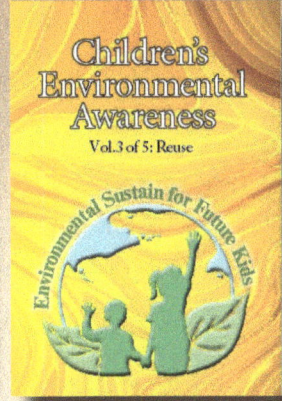

Children's Environmental Awareness
Book Series: Vol.1-5

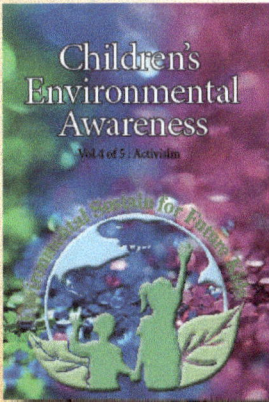

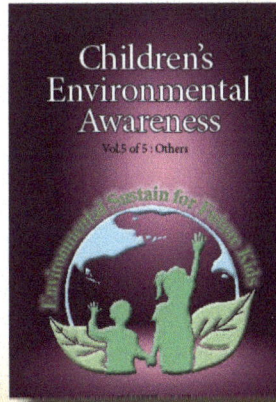

Children's
Environmental
Awareness
Vol.4 of 5 : Activism

Environmental Sustain for Future Kids

Children's
Environmental
Awareness
Vol.5 of 5 : Others

Environmental Sustain for Future Kids

How to Contact us:

Tel.: +1- (778) 751- 8127

Email: info@silosa.ca

Web: http://silosa.ca

Social Media:

Twitter.com/silosagroup

linkedin.com/company/silosa-consulting

https://facebook.com/silosagroup

http://pinterest.com/silosagroup

Telegram: @silosaconsulting

Whatsapp.=1-778-751-8127

www.ingramcontent.com/pod-product-compliance
Lightning Source LLC
Chambersburg PA
CBHW040910210326
41597CB00029B/5040